ADHD
Made Simple

P ADHD

poultonadhd.com.au

By Dr. Alison Poulton
poultonadhd.com.au

What Others Are Saying About
ADHD Made Simple:

"As Designers, we are very pleased and proud to be part of Dr. Poulton's work and research, creating graphic solutions that will help the ADHD community."

~ Gabriel Hidalgo and Edmond Yang

"Wow! Well done. Written by a very experienced paediatrician with many research papers to her name, this book gives a clear and succinct explanation of the medical model of ADHD and will be very useful for any non-medical person supporting a young person, or adult diagnosed with ADHD. It is thought provoking, clear and succinct, and it is a real strength is that there is a supporting website."

~ Dr Anne Fry

Practitioner Psychologist

"As someone who has, for over three decades, followed the large and complex research literature on ADHD, I know first-hand how difficult it is to distil that knowledge in a simple yet accurate manner. Can we achieve that goal when teaching about ADHD? Dr. Poulton's answer is "yes", and her new book shows that she has achieved that goal."

~ Stephen D. Faraone, PhD

President, World Federation of ADHD

"Congratulations. I really enjoyed reading this book and I am sure it will be well received by families, as well as professionals. I also felt that teenagers and youth with ADHD would find the content really valuable. Easy reading, fun visual prompts, practical explanations and strategies to understand and enhance every day over the life span of ADHD."

~ Professor Desiree Silva

Paediatrician and author of the *ADHD Go To Guide*

"*ADHD Made Simple* is a great book for anyone wanting to learn more about ADHD in easy-to-understand terms. It is refreshing to see a doctor providing useful information without using complex medical terminology. In this book, Dr Alison Poulton covers the different traits of ADHD across the ages, along with other co-existing conditions, and not only covers medication that works but also non-medication strategies that are beneficial as well. As a coach, it will definitely be a book that I recommend to my clients."

~ Paula Burgess

ADHD Coach & Author

Illustrations by Edmond Yang (edmond.yn@gmail.com)
Graphic Design by Gabriel Hidalgo (www.gabrielhidalgo.com.au)

Figures 1, 2 and 3 are adapted from:
Poulton A, Nanan R. Australasian Psychiatry 2014; 22(2): 174-8.

Figures 4, 5 and 6 are reproduced from:
ADHD – New Directions in Diagnosis and Treatment. (Norvilitis JM editor)
Intechopen.com 2015. ISBN: 978-953-51-2166-4.
http://www.intechopen.com/download/pdf/48642

Table of Contents

Foreword

As someone who has, for over three decades, followed the large and complex research literature on ADHD, I know first-hand how difficult it is to distil that knowledge in a simple yet accurate manner. Such a task requires us to follow Einstein's maxim that "Everything should be made as simple as possible, but not simpler". Can we achieve that goal when teaching about ADHD? Dr. Poulton's answer is "yes", and her new book shows that she has achieved that goal.

In less than one hundred pages, the happy reader will find clear and accurate descriptions of many of the key findings about ADHD with a welcome focus on understanding ADHD as a neurodevelopmental disorder associated with executive functioning and reward system deficits. Her use of simple examples and cartoons will help lay readers absorb complex concepts and be entertained at the same time. Moreover, rather than being a dry compendium of information, this book does a good job translating information into ideas that patients and families can use to mitigate the effects of ADHD in their daily lives.

Another welcome feature of the book is how Dr. Poulton weaves her clinical experience and wisdom into the fabric of the knowledgebase about ADHD. Because of that, there is much information in the book that will be new, even for those that have read much about the disorder. Her novel ideas about reward, achievement and mood provide a useful framework to help patients, families and clinicians conceptualize the disorder in a manner that should help improve the quality of life for people with ADHD.

Stephen V. Faraone, PhD

Distinguished Professor and Vice Chair for Research
Department of Psychiatry
SUNY Upstate Medical University, Syracuse, NY

Preface

Human behaviour is complex and can be difficult to understand. Adding ADHD into the picture brings yet another level of complexity. Sometimes behaviour may appear completely irrational. But if you look hard enough, there is often an explanation that can make sense of it. In *ADHD Made Simple* I have tried to make the complexities more understandable with the use of examples and pictures.

I have been working with people with ADHD and their families for more than 20 years. I have learned a great deal by hearing about the difficulties that people experience due to their own ADHD or to ADHD in a family member. I have noticed that there are patterns of behaviour that keep recurring.

Parents want to know why their child with ADHD behaves differently from other children. Partners of adults with ADHD also want to understand. The need for good explanations becomes more pressing each time a similar behaviour pattern is described.

Over the years I have considered various explanations and tried them out with my patients and their parents, to see whether they fit with their lived experience. Some explanations have been discarded. But sometimes I see that 'lightbulb' moment when suddenly the behaviour starts to make sense. From understanding develops empathy for the child's point of view, as the parent starts to see reasons behind the everyday challenges. The explanations that 'work' have been further developed, with new insights being added over time.

This book includes some background information on ADHD, with references provided for the relevant chapters. But much of it is unreferenced as it simply reflects my own observations, interpretations and clinical approach. As well as the text, there are also pictures and summary boxes to help those who may feel daunted at the prospect of reading an excessive number of words. For those who may want a more academic version I would recommend my open access chapter: http://www.intechopen.com/download/pdf/48642

I hope that people with ADHD and their families, teachers and therapists will find these explanations helpful and some of the mysteries of ADHD made simple.

1. Introduction – What is ADHD?

Attention Deficit Hyperactivity Disorder (ADHD) is a common condition which may affect 2.6-11% of children and 2.8-4.7% of adults. It is recognised more often in boys than girls, and in men than women. It is a condition in which the mind does not function as efficiently as it should; people with ADHD have impaired functioning due to difficulties with sustaining attention and controlling impulsive behaviour. They also often have a high level of physical activity.

There is evidence that individuals with ADHD have functional deficits in the front part of the brain (the prefrontal cortex), which is the part of the brain most involved in thinking; otherwise called "executive functioning" (see the box below for some examples of executive functioning). ADHD is also associated with reduced motivation for work and activity. This may be due to differences in the way deeper structures in the brain (for example, the striatum and amygdala) respond to rewarding and enjoyable experiences. These emotional responses are affected by the amount of activity in the dopamine reward pathway, which connects the different structures.

Executive functioning:
the functions of the 'thinking' brain

- Reasoning – thinking logically
- Making good decisions
- Short term memory
- Attention span
- Ability to listen and follow instructions
- Controlling impulsive behaviour

ADHD as a neurodevelopmental disorder associated with executive functioning deficits

ADHD is a neurodevelopmental disorder. This means that in ADHD the brain develops and functions slightly differently. ADHD is strongly inherited, and therefore most people with ADHD will be able to think of other people in their family who either have (or might have) ADHD.

It can often be traced back through the generations, although it may not always have been diagnosed, particularly in older people.

People with ADHD often develop strategies to get them through their daily routine that compensate for these deficits. This may make their ADHD less obvious to other people but does not change the underlying brain characteristics.

Although behavioural and self-regulation strategies may be very helpful for people with ADHD, only medication can directly increase the amount of activity in the neural pathways and improve the efficiency of the connections between the different parts of the brain. The stimulant medications used in the treatment of ADHD increase the levels of the neurotransmitters noradrenaline and dopamine, and lead to improvement in the executive functioning (thought processes) and the mood.

How is ADHD diagnosed?

Although formal testing for executive functioning deficits is possible, such tests are not strictly necessary for making a diagnosis. However, a formal diagnosis of ADHD is usually made in accordance with specific diagnostic criteria, such as those published by the American Psychiatric Association in their Diagnostic and Statistical Manual of Mental Disorders (DSM), the current edition being DSM-5.

ADHD is a clinical diagnosis and depends on a person showing the characteristic behaviour to a greater extent than would be expected for a person of their age or developmental level. This behaviour must also be causing problems in their daily functioning; the features of ADHD also occur within the normal population, but not with sufficient severity to cause significant impairment (see picture on page 53).

ADHD typically results in difficulty completing tasks, which leads to underachievement and a less rewarding experience. A popular misconception is that individuals with ADHD will tend to be similar in their behaviour. However, within the normal population there is substantial variation in personality types, skills and abilities. When a diagnosis such as ADHD is added on, it adds a further source of variability. ADHD can occur in people of all levels of intellectual ability. Intellectual ability is one of the most important personal attributes that modifies how well a person manages and copes with their ADHD.

The diagnosis is based on meeting a sufficient number of the DSM-5 criteria for the 3 main types of symptoms or behavioural characteristics – inattention, hyperactivity and impulsivity.

Inattention
People with ADHD have more difficulty for tasks that involve sustained concentration, particularly if the task is mentally demanding. If a person with ADHD is going to complete a task, that task should either be short and easy or be sufficiently interesting, enjoyable or rewarding to keep them engaged. People with ADHD may be able to concentrate for a long time on electronic games. These typically do not involve much independent or creative thought and provide constant stimulation that catches and keeps the attention.

One characteristic of people with ADHD is that they are easily distracted. This can occur while they are talking and may lead to forgetting what they were going to say or losing the point while telling a story. Alternatively, becoming distracted during a task and then forgetting to go back and get it finished can lead to a person being inefficient and disorganised. People with ADHD often have difficulty ignoring distractions and this may make them particularly intolerant to background noise while trying to concentrate (see picture on page 19). Losing focus on schoolwork may lead to disruptive behaviour in class as a response to the boredom that comes with having nothing to do. Lack of attention predisposes to missing instructions and making careless mistakes. A child with ADHD may have difficulty with age appropriate play, quickly losing concentration and moving on to the next task or looking around for something more entertaining to relieve boredom.

ADHD is more disabling in children who have learning difficulties. This is because they must concentrate longer and harder to acquire the same skills. The more difficult the task is for them, the more quickly they will fatigue mentally and give up. Conversely, an able child with ADHD may have no difficulty achieving at school during the early years. However, as the work becomes more demanding in high school, intellectual ability by itself may no longer be enough; if they are unable to concentrate in class and study consistently, their grades may decline. Once a person leaves school, they usually have more opportunity to follow their interests and strengths and ADHD may therefore be less of a problem. However, lack of organisational ability may become more disabling when an individual must contend with the complexities of functioning in society as an adult.

People with ADHD may be easily distracted from tasks that require sustained mental effort

Hyperactivity

Hyperactivity is common in ADHD and is the most easily recognised feature. The hyperactivity often reflects the changing focus of attention as a person moves rapidly from one distraction to another. This restless energy may make it difficult for a person to remain seated for long enough to watch a movie, or their constant tapping or fidgeting may disturb other people. Someone with ADHD may also be excessively talkative, sometimes apparently talking just for the sake of it, and may lack the patience to stop talking and listen.

Hyperactivity tends to lessen with age, and although some adults with ADHD are still hyperactive, others may become underactive and unmotivated.

Impulsivity

People with ADHD often have quick reactions, making decisions without having time to stop, think and consider the consequences. This may include overlooking danger and taking unnecessary risks. Impulsivity can have an adverse effect on peer relationships as a person may unintentionally hurt or offend or repeatedly get scorned for the same misdemeanour.

Impulsive people may lack the patience to wait for their turn or may constantly interrupt. Because of the lack of any active decision-making, these unregulated actions may be considered accidental by the person, who may be inclined to deny responsibility. This lack of impulse control can lead to anxiety and low self-esteem, as the person may suddenly be in trouble without meaning to, and without knowing what they were about to do or say.

ADHD is classified as *combined type* (meets sufficient criteria for inattention and for hyperactivity–impulsivity), *predominantly inattentive* (meets criteria for inattention but not for hyperactivity–impulsivity), or *hyperactive–impulsive* (does not meet criteria for inattention).

The diagnosis of hyperactive–impulsive ADHD tends only to be made in preschool children, who are at a stage of life in which the lack of ability to sustain concentration may be less evident. The symptoms of ADHD may change over the different stages of life, which means that a hyperactive–impulsive young child may no longer be hyperactive in adolescence or adulthood and may have their diagnosis revised to ADHD combined-type, or may not even still meet the diagnostic criteria for ADHD. Does this mean that a person can recover from ADHD? Or does it mean that there is something more fundamental about the way that the brain functions in ADHD that is not always captured by the diagnostic criteria?

ADHD and mental efficiency

In ADHD the various executive functioning deficits may be considered to have the overall effect of making the brain work less efficiently. This means that the more difficult or high-level mental tasks that involve a lot of thought may require an unmanageable level of effort. This is like a runner who has to run uphill. It is not that running is too difficult for him, but he will tire more quickly than others who are running along level ground. He will either keep going but run more slowly, or he will try and run as fast as the others and then have to stop to rest. It is like this for mental tasks for people with ADHD. A task may not be beyond their ability, but it requires a disproportionate – even a super-human – level of effort. This makes it far more difficult for a person with ADHD to achieve their potential. The mental fatigue is genuine and may affect academic functioning, social interactions and managing the daily routine at home.

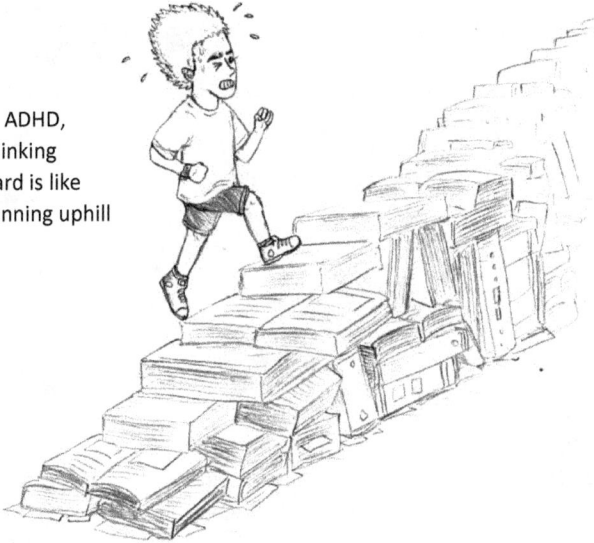

In ADHD, thinking hard is like running uphill

Because ADHD particularly affects the high-order thought processes, the most fundamental problem in ADHD is the inability to achieve consistently at a level appropriate to a person's ability. This includes making difficult decisions when the best choice is not obvious or working out the answer to a problem.

If high-order thinking requires a disproportionate level of effort, a person with ADHD who has to think hard will quickly tire. One way to picture this is to imagine the brain as having a battery that is running out of charge.

The brain in ADHD is like a computer that's running out of charge.

If it runs low-level programs like playing games, it can keep going.

If it tries to run a high-level program like doing homework, it quickly flicks onto standby.

What might happen if the brain's 'battery power' is critically low?	How might this appear in a person with ADHD?

- It may keep flicking on and off   ⟶ • Brief lapses in attention

- It may keep running, but not at full capacity   ⟶ • Person is partially concentrating, but is not taking everything in

- It may work more and more slowly   ⟶ • Increasing effort to keep working on a task

- It may stop working   ⟶ • Daydreaming

Dr. Alison Poulton
poultonadhd.com.au

Losing 'brain power'

As the reserves of 'brain power' are used up, the following may happen:

- The level of attention available for the task reduces to under 50%
- The mind thinks more slowly
- Tasks appear more boring
- It requires an ever-increasing effort to keep going on a task
- Enjoyable distractions become more attractive
- Negative distractions (like noises or people talking) become more annoying
- Information is processed less efficiently
- Information is easily forgotten
- Lapses in attention become longer and more frequent
- It becomes more and more difficult to get started again
- Tasks are left unfinished

NOISE

A person whose brain's 'battery power' or 'brain power' is critically low may learn to adapt in ways that maximise functioning. This may involve conserving power when possible and also ramping up the pressures to keep the brain going when really necessary.

A child with ADHD may rush to get work finished within a timespan for which they can concentrate. Alternatively, they may work for a bit and then stop working and appear to daydream, as if their mind is going blank like a computer on standby. Some just limit their rate of mental effort to a manageable level by working slowly. This may be disguised by giving too much attention to neatness and therefore doing very little of the more cognitively demanding aspects of the work. Creating a distraction may also be an effective work avoidance strategy. For example, a little girl developed the pattern of turning around and giving her mother a cuddle whenever she felt too much pressure to concentrate on her homework. Other more common avoidance strategies include changing the subject or asking an irrelevant question.

Conversation demands mental effort, both for listening and for thinking and formulating the sentences required for a response. Children with ADHD often use strategies which conserve their mental effort. If a child is asked about who they have played with at school, this involves the effort of thinking back to an earlier part of the day and it may be easier to respond with 'I don't remember'.

Children with ADHD often have difficulty carrying out instructions, particularly if given several together. A child may try to look as though he is listening, keeping his eyes on the speaker but not fully concentrating and therefore unable to follow an explanation or instruction. Sometimes a child may only listen to part of a sentence and guess the rest. Remembering several instructions often involves the effort of repeating them mentally. Rehearsal strategies and recall may be less efficient in ADHD. If a person is not putting in adequate mental effort or is distracted by other thoughts, instructions may easily be forgotten.

If a person is achieving less on account of the disproportionate or unsustainable effort they must put into completing a task, they will experience less satisfaction. They may be less ready to put further effort into the next task, with a tendency to give up easily. Inefficient mental processes therefore contribute to the underachievement associated with ADHD and consequent low self-esteem. Some individuals attempt to preserve their self-esteem by reducing their goals in life to a level that is more achievable. This may lead to dropping out of school into unskilled work or state benefits. This may be framed as a deliberate choice rather than failure to achieve.

Children with ADHD often develop various ways of disguising or adapting to the effort they must put into daily life. Some of these could be considered 'taking mental shortcuts'.

2. Conserving 'brain power' and taking 'mental shortcuts'

Avoiding unnecessary effort and conserving 'brain power' is a normal part of everyday life which helps people to function efficiently. This goes for people with or without ADHD. A daily routine involves a series of tasks. If a person's job involves tasks that require intense concentration, afterwards they may feel the need to take a rest or to move on to doing something easier. Similarly, a student who continues studying for too long without taking breaks may become less efficient. This mental fatigue is experienced much sooner in people with ADHD and it may be present before they even start a task.

A person's brain is constantly taking in, processing and analysing external data from their surroundings. This requires effort - or 'brain power'. Different types of data may be easier or harder to process. Visual information may be easier than language-based data, and listening is usually easier than reading. It makes economic sense for a person with ADHD to conserve their precious 'brain power' so that they can keep their brain running for as long as possible. One way of doing this is to get the information from their pre-existing bank of knowledge, as this avoids the need for concentrating and processing new information.

Carrying out a task with the minimum level of 'brain power'

Easy or repetitive tasks that require little active thought, or just involve responding to familiar situations will conserve 'brain power'. However, if a person is not using their brain to its full capacity, this may mean that they fail to notice the additional tasks that need to be done and may therefore lack initiative at work. They may also find it difficult to learn new tasks, as this requires a higher level of concentration.

A common example of a daily routine that conserves 'brain power' is the person (with or without ADHD) who says that their car 'knows' the way home from work. The higher order thinking involved in planning the route for this journey is unnecessary. Attention is only needed for the easier tasks of responding to the traffic lights and to the other road users. 'Brain power' is conserved, allowing the brain to take a rest, or to divert to thinking about other things or concentrating on a radio program. The downside of this is that changes to the normal routine may easily be overlooked, such as making a necessary detour to the shop. However, in ADHD when the person's 'brain power' is critically low, these conservation strategies are a vital part of keeping going. As a result, such oversights are much more common.

Having a regular routine is a good way of conserving mental effort. If you know that as soon as you finish task A it is time to start on task B, you do not have to put any effort into thinking of what to do next. A person with ADHD may therefore manage better with a routine.

'Mental shortcuts'

Inadequate listening

Inadequate listening is probably the commonest type of 'mental shortcut'. This may involve listening to only part of a sentence and guessing the rest or guessing the entire sentence from the situation and tone of voice. This strategy can work well if an instruction is being repeated multiple times, or if the routine is always the same. For example, a child may know that at a particular time in the morning, when a parent shouts out an instruction it is time to go and put their shoes on. However, if the information presented is new or different from usual, it may result in the wrong task being done, leading to inefficiency and frustration.

Listening to part of a sentence and guessing the rest

Doing the bare minimum

If a task must be done, a person with ADHD may look for ways of doing it as quickly and easily as possible. Therefore, a student writing an essay with a word-count may aim to write only just enough words, keeping each sentence as short as possible. Similarly, an adult may have low standards of work, perhaps leaving off the finishing touches or the cleaning up afterwards. Often the last part is rushed, just to get the job finished.

Emotional decision making

Making a good decision often involves high-order thinking and planning. Alternatively, a decision may be made quickly and easily, based on emotion. Emotional decision making therefore conserves mental effort and may be considered a 'mental shortcut' (see page 34).

Conserving effort by talking instead of listening

Some people with ADHD are excessively talkative. Talking too much may be part of being hyperactive. However, for many people with ADHD, talking is easier than listening, particularly if they can talk without putting much thought into what they are saying.

The downside of this is that their speech may be boring, confused or repetitive. It may also prevent meaningful conversation involving talking, listening and responding to another person. Therefore, a person with ADHD may have great difficulty socialising and maintaining friendships.

Procrastination

Procrastination means putting off tasks that require effort. People with ADHD are particularly susceptible to procrastination because they may easily become overwhelmed at the amount of effort that a task will take. This is a bit like a person looking at a steep mountain and realising that they will never be able to climb up and reach the top. Although procrastination may look like laziness, the person with ADHD may, in fact, be a making a realistic assessment of their reserves of 'brain power'.

A person at work who procrastinates may choose to do the easiest or more interesting tasks first because these appear more achievable. These tasks may take a long time as they try to put off doing the harder tasks for as long as possible. By the time the easy tasks are done, the person may have run out of time, or run out of 'brain power', so that the more mentally demanding tasks never get done. This makes the person an inefficient worker.

3. Keeping the brain functioning

If a person with ADHD is going to complete a task, that task has to either be sufficiently short, easy or rewarding for them to be able to sustain their effort.

Tasks that can be done if you have ADHD	Tasks that can't be done if you have ADHD
• Short • Easy • Interesting	• Long • Difficult • Boring

In practice, this is a balance which also depends on a person's natural ability. For example, a bright primary school child with ADHD may find the schoolwork easy and be able to complete it quickly and within a limited attention span. A child who has a particular ability in maths may be happy to do their maths but may find the writing tasks boring and arduous.

Getting up, getting dressed and ready for school or work might not appear to require very much mental effort but for some people with ADHD, even getting through these mundane tasks of life can be difficult.

Maintaining motivation using a reward

Sometimes it helps to structure the day and to plan to have certain tasks completed by particular times, followed by a reward. For example, a person might plan to have each task completed by a particular time and if successful, to follow this with a reward such as watching a television show. This also gives the brain a bit of time to recover before attempting the next task.

Some people find motivation very difficult without external structure, such as a job that has deadlines. A person may actually need the panic of an imminent deadline with negative consequences to be able to get on with concentrating on a task. If a person with ADHD also suffers from anxiety, the anxiety may help the person to put more effort into a task as they worry about what may happen if they do not get it done.

Maintaining attention by increasing the stimulation

If someone has had an attention lapse, they may find it very difficult to recover their line of thought. These lapses interfere with planning and getting through the tasks of the day. Similarly, if someone gets distracted and leaves a task incomplete, it requires an effort to think back to try to recall what they were

Dr. Alison Poulton
poultonadhd.com.au

doing previously. Sometimes they may forget altogether and keep leaving tasks unfinished.

Increasing the level of stimulation may help a person with ADHD sustain their attention. Sometimes the stimulation of music playing may help a person to stay alert. Some people find that speaking their thoughts makes it easier to keep track of them. They may therefore keep up a constant commentary about what they are doing and planning to do next. Although it may help a person with ADHD if they are able to hear their thoughts, this may be a nuisance to other people.

Keeping track of thoughts by thinking out loud.

4. Decision making

Decisions may be rational, emotional, or a combination of both. Making a rational decision involves weighing up the value of the different possible choices. Rational decision-making may also involve taking a long-term view of the decision and its effects on future plans. A decision based on emotion is usually quicker and easier to make and will tend to favour choices that give immediate pleasure. A rational decision may have an emotional component, but this only comes into play for helping to decide between the possible rational options, or when the rational and the emotional choices coincide. In rational decision making the rational mind therefore takes precedence over the emotions.

A person who is saving up for a car may need a new pair of shoes. A rational decision would be to choose the cheaper pair that will serve the purpose.

An emotional decision would be to choose the really nice pair of shoes that is more than double the price. To make a rational decision involves high-order thinking and prioritising as well as exerting mental control over the emotions. This can be particularly difficult for a person with ADHD. However, making a quick decision based on emotion is far easier as it avoids the need to think. In practice, most decisions will have a rational and an emotional component. Therefore, if there are two pairs of shoes that look equally sturdy, it could be reasonable to allow emotion to select the one that is slightly more expensive but looks better. However, a person with ADHD might simply go with the emotion and choose the most expensive shoes, avoiding the cognitive decision-making process, or might even get side- tracked into spending the money on something unnecessary and completely different.

Rational decision-making

Emotional decision-making
(more likely in ADHD)

This pair will serve
the purpose.
They are sturdy, and
cost half as much.
That helps me save up
for a car.

$$

$$$

I MUST
have these
shoes!

These shoes may not be
very special, but I was
able to save some money.
And they actually look
quite nice!

I am so happy
with these
shoes! They
make me feel
so good!

After the decision: an emotional commitment to the outcome
establishes the decision and makes it harder (more effort) to reverse.

This example of emotional decision-making shows a person making a choice that gives immediate, short- term enjoyment, rather than thinking and saving up for a different, delayed, longer-term goal. Decisions which are based on emotion would look impulsive, as they are made quickly and without taking all the consequences into consideration.

After making a decision, a person would want to feel happy with their choice. They may therefore reinforce to themselves why this was the better choice. The person who has made the emotional decision and chosen the impractical, expensive shoes might feel that they had 'fallen in love' with the shoes and could not have made any other choice. The person who has chosen the cheaper shoes would look at them to find their positive attributes. After making a decision, a person may resist re-considering that decision, particularly if the decision was made with difficulty or the person has a high level of emotional commitment to the decision. To reconsider or to go back on the decision might involve considerable mental effort and therefore be particularly difficult for a person with ADHD. The person may therefore get irritated if their partner suggests that the expensive shoes should be returned and exchanged for the cheaper shoes.

5. Living with ADHD across the lifespan

ADHD can be milder or more severe. People with more severe ADHD are likely to run into problems at an earlier stage of life. People with milder ADHD and people who are better able to adapt or learn strategies that help them to cope may be diagnosed later or not at all. ADHD is being increasingly recognised in the elderly, when the aging process not only exacerbates the ADHD but also erodes a person's ability to compensate.

At any age, if a person is not achieving their potential it is important to consider whether this might be due to ADHD. The later a person is diagnosed and treated, the more they are likely to regret the years they have spent struggling and under-performing.

ADHD in childhood

A hyperactive child who is incapable of sitting down and concentrating for more than a few minutes will easily be diagnosed with ADHD. A bright child with ADHD may have no difficulty achieving at school during the early years. However, as the work becomes more demanding in high school, intellectual ability by itself may no longer be sufficient, and if the child is unable to concentrate adequately in class and study consistently their grades may decline. People with ADHD who are exceptionally intelligent and ambitious may find that their ADHD only starts to hold them back once they reach university.

As children mature, they usually develop more control over their behaviour, and this may reduce their reliance on medication. For example, a young hyperactive child may generate so much stress in the family that they may need medication every day. As the child matures the hyperactivity may start to settle and medication may only be needed for school. School is often the most difficult stage of life for a person with ADHD. This is because schoolwork involves prolonged periods of concentration and many of the tasks may not be sufficiently interesting. Once a person is no longer studying, they may be able to cease medication.

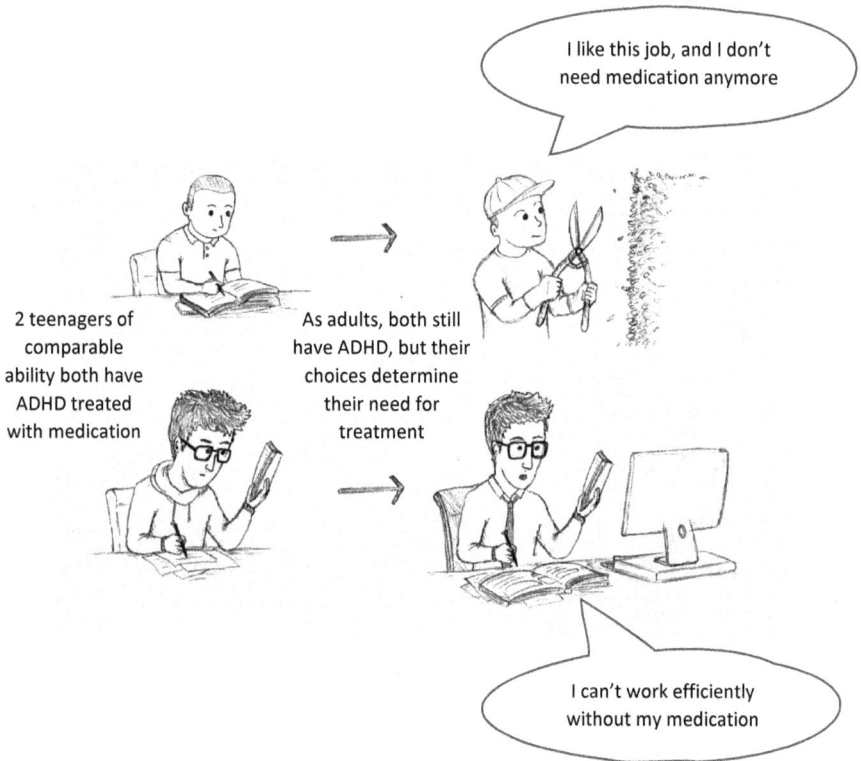

I like this job, and I don't need medication anymore

2 teenagers of comparable ability both have ADHD treated with medication

As adults, both still have ADHD, but their choices determine their need for treatment

I can't work efficiently without my medication

ADHD in adulthood

ADHD used to be considered a condition of childhood, with most people outgrowing their ADHD as they approached adult life. This may be because once they leave school a person has more freedom to choose an occupation that matches their interests and abilities. However, if high-order thinking is unsustainable, they might to choose a job that requires less training or involves tasks that may be accomplished quickly or easily. They might leave the higher-order thinking to their boss who gives them instructions.

ADHD is more often diagnosed in boys than girls, although the numbers are becoming more equal as more inattentive girls are being recognised. However, the years of under-recognition of ADHD in girls means that are substantial numbers of women whose ADHD was missed while they were growing up. Life as a mother in charge of a family usually involves a lot of complex organisation, from running the household and managing the family budget and paying the bills, to organising and remembering all the children's school, recreational and social commitments. Having a job, being a single parent, or having a child with medical problems or ADHD adds to the difficulties. A woman with ADHD is likely to have more problems coping with all the complexities of life and may develop anxiety or depression.

It is important to recognise and treat ADHD, so that people who are struggling can receive the help they need to achieve their goals in life. A person's goals should determine the level of function that they need from their brain, not vice versa. Therefore, the young man in the top picture on page 38 goes into landscaping because this is what he wants to do, not because his poor concentration prevented him from getting the marks he needed for studying engineering. He can relax part of his brain and can still do a good job of pruning the hedge. He may be more creative when he is not on medication. The young man with the desk job must concentrate harder and more consistently. He needs his brain to function at full capacity for the whole day, therefore he continues to take medication.

ADHD in the elderly

The executive functioning deficits associated with ADHD are lifelong. With maturity people often become better at developing strategies to help them to function. However, the physical and mental decline that happens in old age affects memory and the high-order thought processes that are already affected by ADHD. People become less energetic, less motivated and less able to carry out complex tasks. Difficulties with attention may be compounded by worsening vision and the accompanying necessity of keeping track of glasses. Deafness may affect communication: a person who has increasing trouble listening and understanding may adapt by talking excessively, leaving less time for the more difficult task of listening. Decision making becomes more difficult and information and instructions are more easily forgotten. As a person ages, they become less able to put mental effort into managing their ADHD. A person with ADHD may have developed good strategies for remembering the tasks of the day, perhaps with lists or using a calendar or diary. However, declining mental functioning not only adds to the problems associated with ADHD, but also affects a person's coping mechanisms. Sometimes ADHD may only become apparent in a person following the illness or death of their partner, who up until then had been doing all the organising and decision-making.

An older person is highly unlikely to have been diagnosed with ADHD in childhood. However, as ADHD is a common condition that is lifelong, it is also likely to be common in old age, particularly when there is ADHD in the family. Older people may respond well to treatment; therefore, it is important not to overlook ADHD in the elderly.

Dr. Alison Poulton
poultonadhd.com.au

6. Oppositional Defiant Disorder (ODD) / Antisocial Personality Disorder

Oppositional Defiant Disorder (ODD) is very frequently associated with ADHD in children. The same is also true in adults, except that ODD is more often called Antisocial Personality Disorder, a condition associated with disregard for the rights of others. However, for the sake of continuity with the childhood disorder, in this book we will be using the term ODD to include all the disruptive, impulse control and conduct disorders (ODD, Intermittent Explosive Disorder, Antisocial Personality Disorder, Disruptive Mood Dysregulation Disorder, Conduct Disorder). These conditions are associated with temper outbursts (problems in emotional regulation) and with behaviour problems including rule breaking and antisocial acts, with the specific diagnosis designated according to the main symptoms and the relative balance of the mood versus the behavioural dysfunction.

A person with ADHD and ODD may misinterpret a situation and overreact with aggression.

People with ODD typically overreact with anger in response to minor frustration. The lack of control over impulsive behaviour in ADHD becomes even worse when associated with anger. A person with ODD may incorrectly interpret another person's actions as hostile and impulsively over-react with aggression. Therefore, people with ODD are at greater risk of committing violent crime.

ODD is also associated with deliberately annoying people and sometimes with planned acts of spite. This can cause massive problems within the family. It is often associated with a negative attitude and a tendency to blame other people and deny that they are at fault. People with ODD are argumentative and may oppose authority and appear negative in their mood and outlook. This can lead to a lot of problems at work.

7. Mental Effort-Reward Imbalances Model (MERIM): Why do people with ADHD so often have ODD as well?

The most fundamental problem in ADHD is the difficulty with concentrating sufficiently to get the necessary tasks finished. This can be thought of as mental inefficiency, with everyday tasks requiring a disproportional effort. For those without ADHD, getting through the daily routine involves a constant stream of tasks that all require some mental effort. Successfully finishing these tasks gives a series of achievements. Although most achievements are small, they are each associated with the satisfaction of task completion – the feeling of a job done well. These small feelings of success all help to sustain a stable and happy mood. Therefore, for example, you get up in the morning, you put some effort into getting dressed and ready to go. You look presentable in your clothes. You have achieved and you feel good about yourself and ready to put further effort into the next challenge.

Achievement therefore involves some level of effort and is associated with a feeling of satisfaction (reward) which contributes to a good mood and a readiness to attempt the next task. Good mood is important for normal functioning. People who are fortunate enough to have a happy disposition are generally well-liked and tend to have better emotional, psychological and social well-being, which means better physical health and fewer days off work. The cycle of achievement, reward, good mood and further achievement is represented in Figure 1.

ADHD Made Simple
7. Mental Effort-Reward Imbalances Model (MERIM): Why do people with ADHD
so often have ODD as well?

Figure 1. Achievement and reward sequence

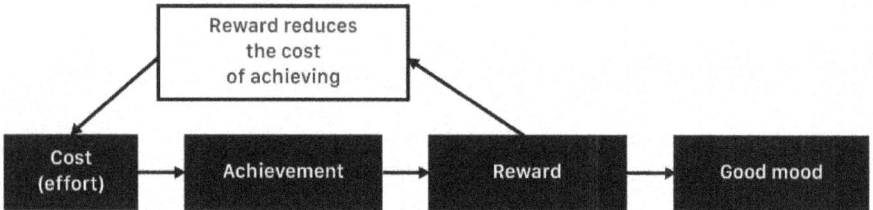

In ADHD, the pathway shown in Figure 1 does not work as effectively as it should. There are two places where there can be problems. The first is if the cost or effort of concentrating well enough to achieve is too great, as occurs in ADHD. The second is if a person does not experience enough satisfaction to make the task worthwhile. This is what happens in ODD. Therefore, people with a combination of ADHD and ODD find achievement doubly difficult. The lower level of successful task completion is associated with lower mood. This is illustrated by the Mental Effort-Reward Imbalances Model (MERIM), shown in Figure 2. and illustrated in the picture on page 45.

Figure 2: Mental Effort-Reward Imbalances Model (MERIM)

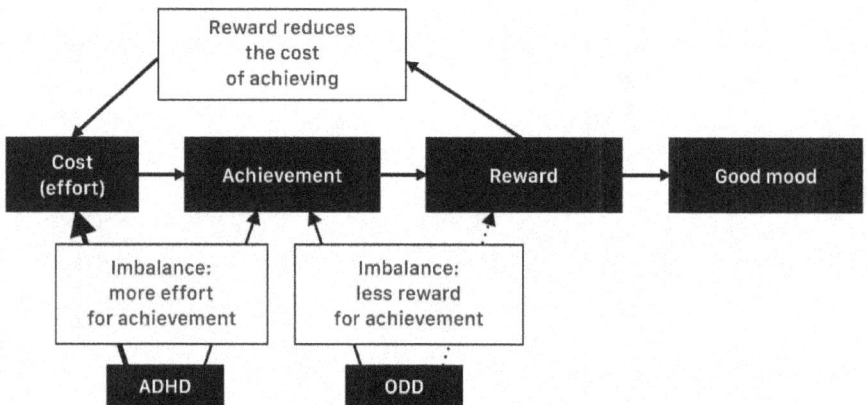

Dr. Alison Poulton
poultonadhd.com.au

ADHD Made Simple
7. Mental Effort-Reward Imbalances Model (MERIM): Why do people with ADHD
so often have ODD as well?

This picture represents a task at school that the boy on the left finds easy and satisfying. The same task looks very different to the boy on the right with ADHD and ODD. The task is not only outrageously difficult, but the reward is tiny.

ADHD+ODD means:
- greater effort
- lower reward

Several of the DSM-5 diagnostic criteria for ADHD are outcome-based and relate to lack of achievement in task completion. These do not dictate the causal mechanism and are therefore not specific to underachievement due to executive functioning deficits. Children who underachieve due to inadequate experience of reward would also qualify. According to the MERIM, the dual deficits in executive functioning and reward experience contribute independently to the lack of achievement associated with ADHD (Figure 2). Because these two mechanisms are both highly prevalent and are additive in their effects on achievement, most people diagnosed with ADHD using the

ADHD Made Simple
7. Mental Effort-Reward Imbalances Model (MERIM): Why do people with ADHD
so often have ODD as well?

DSM are likely to have some degree of deficit in each. This provides an explanation for the substantial levels of diagnosable ODD among children with DSM-diagnosed ADHD. The MERIM would therefore consider the negative attitude and outlook that is frequently associated with ADHD as evidence of some degree of reward deficiency syndrome contributing to the symptoms of ADHD.

Inadequate experience of reward leading to symptoms of ODD

Getting pleasure from the little things in life is important as this helps to maintain a good mood and amicable outlook. However, if the subjective experience of reward is low, a person is likely to feel negative and dissatisfied. This low mood may lead to poor motivation – the feeling that a task is not worth the effort. Alternatively, a person may compensate by seeking activities that are more highly rewarding or that give reward for less effort. The dissatisfaction associated with inadequate experience of reward predisposes to the characteristic behaviour of ODD.

People who have deficits in their experience of reward may feel miserable and moody – their anger is nearer the surface. They may also compensate by seeking activities that are more highly rewarding or that give reward for less effort. These rewarding activities make them feel happier. In ODD the reward-seeking behaviour is maladaptive and destructive.

When people communicate, they have an effect on each other. In a conversation, you would feel you had communicated successfully if the other person appeared interested. This might make you feel good about yourself. Making a person feel happy is even more rewarding. But it is not always easy to tell a good joke that makes people laugh or give someone a pleasant surprise that makes them happy. It is often easier to affect someone's emotions by irritating, hurting or upsetting them. This is what happens in ODD.

ADHD Made Simple
7. Mental Effort-Reward Imbalances Model (MERIM): Why do people with ADHD
so often have ODD as well?

The big rewards are the social rewards

If you make someone happy, you feel really good.

But it's often easier and more reliable to make someone angry or unhappy. This is what happens in ODD.

People with ODD are often argumentative, may deliberately annoy other people and may be spiteful or vindictive. Although these strategies might not appear to be rewarding or enjoyable, it is difficult to imagine a person being deliberately spiteful if this were not pleasurable in some way. A playground bully would not be a bully if they did not enjoy bullying.

A person who does not experience enough reward may feel a bit better after being deliberately difficult or upsetting someone. This behaviour, which compensates for deficits in the subjective experience of reward, may be addressing a problem with their brain chemistry. The behaviour therefore works for the person with ODD but clearly does not work for their family and workmates. The positive impact of bullying on a person's mood is shown in Figure 3.

ADHD Made Simple
7. Mental Effort-Reward Imbalances Model (MERIM): Why do people with ADHD
so often have ODD as well?

Figure 3: Oppositional behaviour compensates for lack of reward

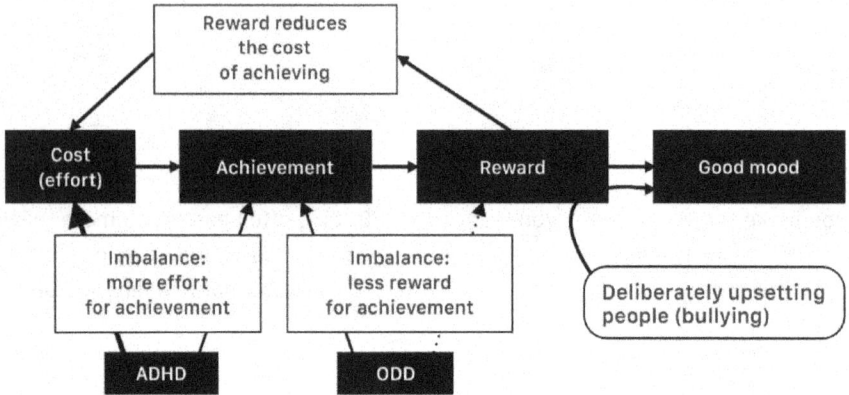

This child has gone to some effort to cause chaos and is quietly enjoying the result.

Parents fighting

Winning is a highly rewarding experience. Some people use competition to maintain motivation by making every activity into a win–lose situation. Such

Dr. Alison Poulton
poultonadhd.com.au

ADHD Made Simple

7. Mental Effort-Reward Imbalances Model (MERIM): Why do people with ADHD
so often have ODD as well?

people may be unable to tolerate losing. Teenagers or adults may actively look for opportunities for starting an argument that they think they can win. The following day they may argue for an opposing view if they think that this will give them the upper hand. Parents sometimes say that their child with ADHD will argue that black is white. Perceptive parents may observe that their child would start out in an angry mood but after a prolonged argument that frustrates or even hurts and upsets the parent, the child's mood may have improved. Alternatively, the child may be skilled at annoying or upsetting other family members, or playing one person off against the other, and experiencing quiet satisfaction at the resulting chaos. With increasing sophistication, the underlying spite may be heavily disguised, and the hostility appear 'accidental'.

It may be useful to consider ODD as a condition with rules that people with ODD tend to follow. This may help to predict a person's likely response.

Rules of Oppositional Defiant Disorder (ODD)

Overriding principle – the need to win

* Never admit to being wrong
* Always disagree
* The answer to any request is "No!"
* Look for every opportunity to get the better of someone
* Winning is more important than reason or fairness
* Try to appear innocent by blaming someone else

One of the biggest problems with ODD is that the need to win an argument and the tendency to deny any fault or weakness may make a person adamantly

ADHD Made Simple
7. Mental Effort-Reward Imbalances Model (MERIM): Why do people with ADHD
so often have ODD as well?

oppose the need for treatment. Even if aggression lands the person in serious trouble, they may still argue and blame the circumstances or other people and deny their own responsibility. Such people can be very difficult to help.

However, being competitive may be used adaptively to enhance the reward associated with routine tasks or chores, for example, a child trying to break their own record for how quickly they can get dressed.

Relating the MERIM to other models of ADHD

The MERIM is not the only model for ADHD that competes with that described in the DSM. The MERIM is probably the simplest model, as it does not attempt to relate the symptoms of ADHD to specific testable modalities of executive functioning. Instead, it starts from the premise that unspecified executive functioning deficits mean that thinking is less efficient in ADHD and therefore requires more mental effort. It also does not really provide any explanation for the hyperactivity or the impulsivity.

It has similarities with the model put forward by Douglas, which considers ADHD to be a result of four predispositions: the desire for immediate gratification, reluctance to invest mental effort, impaired impulse control and impairment in modulating arousal or alertness. However, although it does not address arousal and impulse control, in the area of gratification the MERIM goes further than Douglas in that the desire for reward is explained as being an intrinsic deficit that reduces the individual's subjective experience of reward and interacts with the motivation for mental effort.

Barkley postulated the primary problem in ADHD to be inadequate response inhibition. He gave this as the underlying cause for deficits in executive functions that include working memory, inner speech and verbal reasoning, analysis of behaviour and for deficits in emotional regulation. According to this model, the symptoms of ODD would be explainable as manifestations of the

ADHD Made Simple
7. Mental Effort-Reward Imbalances Model (MERIM): Why do people with ADHD
so often have ODD as well?

emotional dysregulation associated with ADHD. Therefore, with Barkley's model, there is also no need for any additional diagnosis of ODD. However, although Barkley's model includes ODD in the overall symptomatology of ADHD, unlike the MERIM it does not explain the observation that the main emotional component of ADHD should be negative.

8. ADHD as a continuum

The features of ADHD and ODD are often said to be continuously distributed throughout the population. This means that while some people clearly do have ADHD and others clearly do not, there are all different shades of grey between the extremes of black and white.

This person clearly has ADHD

In between are all shades of grey

This person clearly does not have ADHD

<< More ADHD   Less ADHD>>

This creates a challenge for diagnosing ADHD, because there is no precise cut-off between those who do and those who do not have ADHD. The same goes for ODD. But the positive side of this is that everyone can understand what it feels like to have difficulty concentrating or to find that a task is just too much effort, because everyone experiences this sometimes. It also means that having an understanding of ADHD and ODD also helps for understanding people in general.

Just as people with ODD may be striving to make their lives feel more rewarding, everyone wants to achieve adequate reward with a manageable level of effort. Most people would expect to get most of their reward from their routine everyday tasks, such as their work, talking with their friends and family (positive social interactions), entertainment and bodily functions such as eating. Within the broad categories of chores, schoolwork and social interactions, different activities will vary in their level of interest and difficulty, with some chores and schoolwork being experienced as more rewarding and less arduous than others.

Figure 4 shows a range of activities that vary in the amount of effort they require and the level of reward experienced. People with ADHD find it difficult to put in the mental effort. People with ODD who experience little satisfaction from the everyday activities of life would tend to seek tasks that are high in reward. If someone has both ADHD and ODD, for an activity to be worthwhile the level of reward has to be particularly high for the level of effort. If the reward is great enough, they may be able to make considerable effort.

Figure 4: Hypothetical balance of effort and reward for various tasks and pastimes

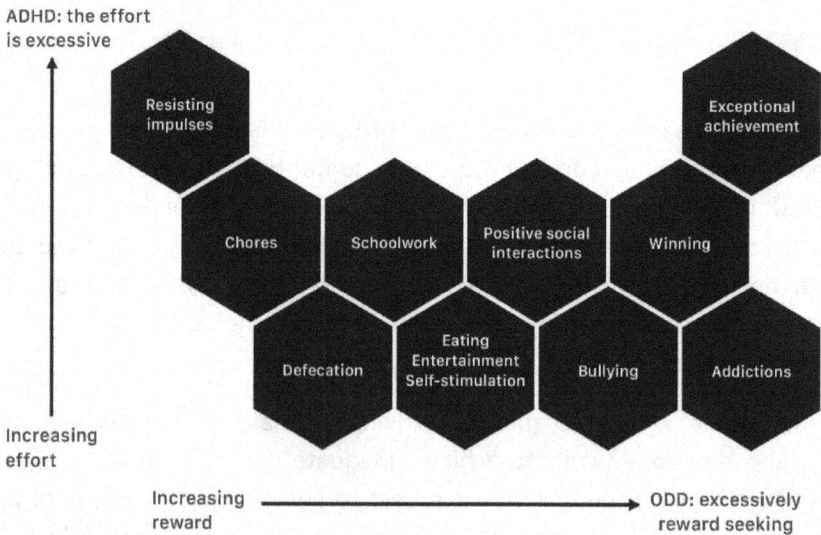

Intellectual capacity and tenacity determine an individual's capability for achievement. Individuals with severe ADHD, ODD and intellectual disability may find that going to the toilet for defecation is not sufficiently rewarding to be worth the effort. An individual may indulge in self-stimulation as this provides reward and is not mentally demanding. Eating is also easy and rewarding, which may explain the recognised association of ADHD and ODD with obesity. People with ADHD and ODD are particularly susceptible to addictions to substances such as nicotine or illicit drugs. Conversely, resisting impulses and resisting emotional decision making requires substantial effort and is not particularly rewarding. An aggressive and irritable child may therefore have no meaningful incentive for putting effort into foregoing the satisfaction of hitting a sibling.

Having a higher level of intellectual ability opens the possibility for higher levels of achievement. Within the broad categories of chores, schoolwork and social interactions, different activities will vary in their level of interest and difficulty for the individual, with some chores and schoolwork being experienced as less arduous and more rewarding than others.

... so after that, she told me ...

Are you actually listening?

Listening requires effort, making conversation difficult

The level of effort required for social interaction is often underestimated. Adolescents and adults generally demand a high level of attention from their friends and even though conversation is rewarding, a person with ADHD may find the intensity of the mental effort unsustainable. A child, for example, may consequently withdraw to a less demanding pastime, perhaps playing alongside their friend. Alternatively, a child with ADHD may be more comfortable playing with a younger or less intellectually demanding child, or an older child who can make allowances or entertain. They may find relaxation from a low-level, repetitive activity, which can lead to an incorrect diagnostic label of autism spectrum disorder.

If a person's experience reward is inadequate, they will tend to feel low and dissatisfied, with their anger easily triggered by minor frustrations. They may also be striving for the higher rewards. Some people with reward deficit may be intensely competitive. Those who are intellectually able may strive for exceptional achievement. If they fail, this may lead to hostility towards those who succeed. Alternatively, they may feel better after they find someone to bully, as happens in ODD.

With the exception of addiction, the higher rewards depend for their value on social recognition or an emotional response from one or more other people. Even exceptional achievement needs a social frame of reference in order to designate its value. The higher rewards associated with more positive achievements tend to require higher levels of effort and aptitude. By contrast, negative behaviour such as bullying has a high balance of reward for the mental effort and is therefore easier for those who are less able. However, the rewards that are associated with low levels of effort and achievement may be associated with low self-esteem. This could negate some of the reward experienced from activities such as bullying. Attributing blame to the victim may reduce this negative effect on the bully's self-esteem.

- Most people get most of their reward from positive social interactions, task completion, entertainment and positive bodily functions.
- For people with reward deficiency/ODD, greater rewards are needed to rectify their low mood. The strategies people use for compensating will depend on other personal attributes – for example exceptional achievement is only possible for those with high ability.
- Gaining an emotional response from other people is highly rewarding, but negative responses (from bullying) may be easier to organise than positive responses.
- A person's mood is a measure of the success of their strategies.

---

9. Diagnosing ADHD: the importance of assessing functional impairment

In most people with ADHD the diagnosis can be made by detailed questioning:

- Symptoms of ADHD (inattention, hyperactivity and impulsivity);

- Problems and difficulties that a person has experienced because of these symptoms (impairment in functioning – see box below);

- Strategies that a person uses that help them to get through their daily routine.

The key to recognising ADHD is not simply a matter of seeing the symptoms but, more importantly, it relates to the resulting problems in functioning. When assessing the extent of functional impairment, it is useful to consider the level of achievement in the following modalities: achievement in relation to ability; peer relationships; ability to function at home and/or at work without generating unreasonable levels of stress or disruption; and level of self-esteem.

ADHD can affect the following areas of functioning:

- Achievement in relation to ability
- Social and working relationships
- Ability to function at home without generating unreasonable levels of stress or disruption
- Ability to function at work without generating unreasonable levels of stress or disruption
- Level of self-esteem

Percentage efficiency

One way of assessing the severity of ADHD is to consider the percentage of time that a person's mind is on task while they are working. For example, a teenager whose attention span is only 10-15 minutes will be concentrating for only 25% of a lesson that lasts for an hour. Unless the teenager is exceptionally able, they are likely to drop further and further behind as the demands of school increase. Similarly, a person who is only able to concentrate for 25% of the working day will never be as productive as they should be.

Baseline mood-setting

When evaluating a person for ODD it is useful to ask about their baseline mood to find out whether it is lower than normal. Some people with ODD rarely appear to be happy and spend a higher percentage of the day being irritable and negative in their attitude. The person with ODD may not be aware of this because they consider their low mood to be normal. The moodiness may often be more readily apparent to the person's partner, other family member or workmate. Assessing the percentage of the time that a person's mood is happy or pleasant can be informative.

Dr. Alison Poulton
poultonadhd.com.au

10. ADHD and mental illness

Deficits in reward do not only occur in ADHD/ODD but are also associated with other conditions such as addictions and obesity. These are sometimes termed 'reward deficiency syndromes', and they are characterised by the strategies that people use to compensate for their inadequate experience of reward. These may include comfort eating, compulsive gambling, internet or gaming addiction and drug abuse.

Mental illnesses, such as depression and anxiety, are common, particularly in people with ADHD. If a person is achieving less due to the unsustainable level of effort they have to put in, they are likely to experience less satisfaction. They will also experience more mental fatigue associated with thinking, or with meeting the intense demands associated with socialising and conversation with friends. They may be less ready to put further effort into the next task, with a tendency to give up easily. Inefficient mental processes therefore contribute to underachievement in ADHD and consequent low self-esteem. This may be associated with depression, anxiety or substance abuse. It is very important for the psychiatrist or therapist to screen carefully for the possibility of underlying ADHD. If there is untreated ADHD, treatment of the associated mental condition is unlikely to be very successful.

Depression

Low self-esteem associated with the difficulty of achieving goals that ought to be manageable, may progress to depression. The inability of friends, family or workmates to comprehend that the difficulty a person with ADHD has with sustained concentration is genuine may make matters worse. Being blamed for laziness or inefficiency may contribute to feelings of low self-worth and depression.

Anxiety

ADHD increases susceptibility to anxiety. If a person's mind is not fully focussed or they have frequent lapses in concentration, important information may either be missed or forgotten. This may lead to frequently being in trouble for not knowing what to do or not doing things correctly. The ever-present fear of failure may be associated with anxiety. Impulsiveness may also lead to anxiety because a person may keep saying or doing the wrong thing without being able to stop, think and consider the outcome.

Anxiety may sometimes mask ADHD when fear of failure increases the motivation to achieve. However, this increased effort has a cost and afterwards the person may be mentally exhausted and short-tempered. Worrying may also be very tiring, adding to the mental fatigue associated with ADHD. Therefore, although a little bit of anxiety may help focus the mind in ADHD, too much anxiety will not.

Obsessive Compulsive Disorder (OCD) is a type of anxiety in which people do repetitive activities as a way of relieving their worrying thoughts. For example, a person may have an irrational fear of germs and keep washing their hands or may keep checking that the house is locked if they are afraid of burglars. OCD may be worse in ADHD because if a person is not concentrating on the tasks of the day, they will have more time to spend on their OCD. The compulsions also act as a further distraction that takes their mind off their tasks. Treatment of ADHD may therefore improve the response to therapy in OCD.

Substance use disorder

People with ADHD have a greatly increased risk of addiction. This is partly due to impulsiveness and difficulty with making good decisions. If a person with ADHD is having difficulty achieving, they may drop out of school or work and get into bad company and start taking drugs. Alternatively, a person may use drugs to make themselves feel better, directly addressing their reward deficit. Addiction to cigarettes is particularly common in ADHD because of the calming effect of the nicotine. Treatment of substance use may be more successful if the ADHD is also treated.

Gaming and Internet addiction

Games are designed to be compelling, entertaining and rewarding. The rewards may be visual, such as seeing a target explode, or may result in accumulating points. It can be very difficult for a person to stop playing and get on with their life, because they are always wanting the next win. Some online games involve other players, which adds negative social consequences for leaving a game unfinished.

People with ADHD often experience more boredom than others because they find the activities that they should be doing require too much effort to be worthwhile. Electronic entertainment that constantly stimulates and rewards and requires no independent, creative thought may be particularly addictive to people with ADHD.

Obesity

Obesity is a complex disorder that usually has a mental component. Eating appetising food is a rewarding activity that does not require mental effort. Eating may therefore be used to address reward deficiency, briefly making the person feel a bit better. This is often called 'comfort eating'. A person with ADHD is likely to have a lot more difficulty with controlling impulsive eating and with putting effort into increasing their level of exercise level in order to lose weight. As a result, ADHD is often associated with obesity.

11. Non-pharmacological management of ADHD

Management of ADHD may involve medication, non-pharmacological treatment or a combination of both.

Non-pharmacological approaches to management usually focus on the areas of functioning which are causing the most problems. This may involve additional learning support or other assistance related to the executive functioning deficits, such as help with organisation. However, the main emphasis is usually on behaviour management strategies. The conventional behavioural strategies used in ADHD are not specific but aim to take good parenting and good classroom management to a higher level. Therefore, strategies may be applied to the whole family or to the entire class or even the whole school. An additional but less well-utilised modality of non-pharmacological management targets the emotional issues. We suggest a larger role for emotional self-regulation as a means of promoting and maintaining a positive mood and outlook in ADHD.

Addressing the imbalance of the level of mental effort required for achievement

Additional learning support
This is designed to address specific problems exacerbated by the executive functioning deficits, for example, additional support with reading, so that this skill becomes easier and accomplished with a more manageable level of effort. Reading is a complex skill that involves several components. Each component requires attention. Therefore, the individual has to recognise the letters, relate them to their sounds and blend the sounds together to decipher the word. The words then have to be remembered so that the sentences can be derived. The

sentences have to be understood and their meaning remembered long enough to make sense of the passage. The main reward of the task is in the interest from the information contained in the passage.

When a child is learning to read, the process is slow and laborious and the reward from the information may be lost unless the sentence is very simple. With practice, the child starts to recognise common words without having to sound out each one individually and reading becomes easier and more fluent. This allows more attention to be focussed on the meaning. The information is received at a faster rate and the balance of effort to reward improves. As reading becomes more rewarding, the child may start to read books for pleasure and thereby further practice and develop their skills.

If someone has ADHD, their attention span for concentrating will be less than other people's. The learning process may be more laborious, and they may be inclined to give up easily. It may be harder for a child with ADHD to attend to the meaning while simultaneously deciphering the individual words, which reduces the interest of the task for the child. At this stage, additional one-to-one teaching may accelerate the rate at which the child develops reading fluency. As reading is necessary for almost every area of schooling, good skills that enable a child to read without putting all their attention and effort into the process is very important.

Modifying the tasks and expectations

A person with ADHD is likely to need a higher level of attention, and the tasks may need to be modified to make them achievable within a person's limited attention span. Such strategies may include keeping tasks short and varied, and moving on to a new topic before boredom sets in. Instructions need to be easily understood and repeated if necessary, perhaps with a written task list. A person with ADHD may need to be reminded to remain on task. Breaks may be factored in, such as sending the person out on an errand.

Teaching organisational strategies
Organisational skills can also be taught. These can include strategies to keep track of work, including structuring the tasks, using checklists and long-term planning of tasks with their completion dates.

Addressing the imbalance in the level of reward experienced: increasing the external rewards with conventional behaviour management

Behaviour management strategies are designed to make favourable behaviour more rewarding and negative behaviour less rewarding. These strategies usually involve a combination of rewarding desired behaviour, and negative consequences for behaviour that is being discouraged. They depend on the child being able to evaluate in advance the consequences of their behaviour. The behavioural strategies must be carefully thought out.

Conventional behaviour management has drawbacks: because the rewards and consequences are external and often tied to particular tasks and situations, they may not carry over to *other* tasks and settings. Conventional behavioural strategies often use emotional rewards, with the parent or teacher praising the child and showing delight if they have achieved or put considerable effort into the task. The child may respond by trying harder in order to gain the satisfaction of making another person happy. Therefore, it is frequently observed that a child will work better for a teacher who cares and takes more interest in them, but then works less well following a change of teacher. The long-term aim of behaviour management is that the behavioural change should become habitual as the child matures.

A review evaluating psychological interventions has demonstrated sufficient evidence to consider behaviour management to be an established and effective intervention for ADHD, either when administered by the parent following training or when used in the classroom. For behaviour management to be effective, a number of prerequisites must be met.

1. The child must be capable of carrying out the target behaviour.

This means that the goals should be realistic. It is important that goals are not too difficult such that the child gives up. Targeting small, manageable tasks is often the more effective approach. In children with a lot of behaviour that is perceived as problematic, goals need to be prioritised. For example, if a child regularly refuses to do any work, rewarding them for concentrating for 5 minutes and writing a single sentence and gradually working up to completing their entire half hour of work may be more successful than choosing 'work completion' as the initial goal.

2. The child must understand the rewards and consequences and be able to relate these to their behaviour.

The child needs to have sufficient capacity to be able to comprehend that there will be consequences. The child also has to make an emotional connection with the consequences. It has been shown that children with ADHD may choose immediate small rewards over larger, delayed rewards. The relevance of this to the clinical setting is that people with ADHD may appear to 'live for the moment'. The child may at be able to recite the consequences for a particular misdemeanour; but at the moment of making a decision, the consequences appear to have little relevance to them. Afterwards, the child may show no interest in the reason for their punishment, experiencing it only as a frustration. It may not be that the child is intellectually incapable of understanding the connection between an activity and its consequence, but that what is important or relevant to the child is the here and now.

Dr. Alison Poulton
poultonadhd.com.au

3. **The rewards and consequences need to be meaningful and appropriate.**

Rewards and consequences should be chosen carefully. A child might be rewarded with time to play on a computer; a meaningful punishment might be taking away the child's favourite game. Rewards and punishments that are small and repeatable are often more effective than larger ones. For example, if a parent is very angry with the child, there may be a temptation to extend the duration of the punishment, perhaps taking away the favourite toy or banning the child from watching television for a week. If the child subsequently misbehaves during that week, the parent has lost one valuable option for punishment. Alternatively, if the punishment is milder, for example, the child is prevented from watching just one show, or loses their game for only five or ten minutes, the same punishment can be repeated as often as necessary. Prolonged punishment with restoration of the item made dependent on good behaviour may be even less effective. To a child with ADHD, a week may be such a long time that they consider the item lost forever; furthermore, it may be unrealistic to expect the child to behave well for a whole week. Withdrawal of attention from a child who has misbehaved can also be effective.

4. **The strategies should be applied consistently.**

Effective behaviour management requires consistent effort from the parent or teacher. If there is any leeway a child may become skilful in picking the time when they can get away with breaking a rule.

5. **The child must choose to cooperate.**

Cooperation is likely to depend on the child's own assessment of the balance of effort to reward. If the effort required is disproportionate due to the executive functioning deficits associated with ADHD, the child may insist on a reward that appears similarly disproportionate. For example, a small reward, such as adding a sticker to a chart for every task completed, may work for a child for a few days until they realise that the stickers are not worth the effort. At that stage, in order for the behaviour management to continue to be effective, a higher reward may be negotiated. This cycle may continue until the child will not even consider doing any homework unless rewarded with a very substantial sum of money.

Alternatively, a child may perceive that they will experience greater satisfaction through non-cooperation. Figure 4 (page 54) categorises bullying – behaviour designed to upset or hurt another person – as being more rewarding than work. Therefore, if a person can derive an alternative to cooperation that causes pain, this may appear an attractive option. If a child perceives that the parent or teacher is emotionally committed to their cooperation and genuinely wants to see the child carry out the task, this may provide an opportunity for bullying. This might take the form of deliberately destroying their work, for example, by scribbling on the page. Observing the resultant surprise, anger or frustration may be immensely satisfying for the child. Another very common strategy for non-cooperation is arguing. This may be a delaying tactic, and a parent may be baffled that their child may spend twenty minutes and considerable effort arguing over ten minutes of work, which ultimately still has to be done. To the child, the arguing may serve several purposes. Firstly, time spent arguing may be considered time well-spent because the work is not actually being done. Secondly, the child may be negotiating a better deal, such as a higher reward or a reward in advance of the task. Winning such a concession would also be rewarding in itself (Figure 4, page 54). Thirdly, the child may be bullying the parent, enjoying the effect of the argument on their parent's emotions, for example, observing an increasing level of frustration or anger. It is important for adults to understand the value that a child may place on observing an emotional response. Withdrawing from the child to calm down may minimise the reward the child experiences for their negative behaviour.

Enhancing reward with emotional self-regulation

Emotional self-regulation with the aim of improving the mood fits in with the logic of the MERIM, because for most individuals, the reward deficit resulting in a less positive mood contributes to the symptoms of ADHD. Although strategies that can lead to a higher level of task completion have merit, an important additional outcome is the effect on mood. People with ADHD and reward deficiency may have their baseline mood set at a lower level than normal, making them feel somewhat irritable for much of the time. Therefore, it would make sense to develop strategies for improving the mood.

If, for example, a child completes homework under protest and with the sole aim of gaining a tangible external reward, perhaps perceived as a bribe, this might be considered an acceptable outcome as the work is done. However, if the attitude towards the work is poor, it is likely that the child will complete it to the lowest acceptable standard. Therefore, an important additional aim would be to teach the child to value their work and gain internal reward in the satisfaction of a job done well. In other words, the positive aspects of the task that has been undertaken would be used to enhance the mood.

Emotional self-regulation could supplement conventional behaviour management based on rewards and punishment, but places less emphasis on targeting particular behaviour, instead focusing on generating a positive mood through achievement.

Individuals may use several strategies which regulate their emotions. These strategies are not simply learned in childhood and adolescence but continue to develop, usually in a positive way, over the course of adult life. Self-regulation strategies may be helpful as a long-term intervention for generating and maintaining positive emotions. For example, an intervention study of meditation (Loving Kindness Meditation) found that 35% of participants continued to derive positive emotional benefit from meditation a year after ceasing therapy.

Unlike behaviour management, which relies on external rewards, emotional self-regulation aiming to promote positive emotions has a theoretical advantage that its techniques may directly address the underlying reward deficit. Furthermore, it can be applied to all aspects of daily life, and once taught and adopted, it does not rely on any outside sources for reward as individuals evaluate and provide their own reinforcement for their positive behaviour, developing strategies for sustaining their mood and self-esteem. The long-term goal would be for the person to become independent in using the techniques of emotional self-regulation. This might happen if the person notices that these strategies are worthwhile because they make him or her feel better. Because the individual has the control, self-regulation in relation to mood promotes individual responsibility and independence.

Positive rumination

Rumination involves repetitive thoughts that can influence an individual's emotional state. Rumination is conventionally considered to be negative, as the repetitive thoughts are distressing and can lead to a range of mental health problems including depression and anxiety by focusing on the symptoms and causes without seeking any solution to the perceived problems. Negative rumination not only exacerbates depression and anxiety but is also a risk factor for a range of mental health problems, including aggressive behaviour in men.

However, we suggest that spending time reflecting on a positive achievement could increase the level of enjoyment or satisfaction obtained from it. This is called positive rumination. It may be a strategy that happy people use to sustain their positive mood and amicable outlook. As it is a cognitive process that would involve some mental effort, it may come less easily to individuals with ADHD.

If a person completes a task with the sole aim of avoiding getting into trouble, this might be considered an acceptable outcome as the work is done. However, if the attitude towards the work is poor, it may be done badly. An example of positive rumination would be for a person to spend a bit of time admiring the good points about a piece of work that they have done and then reflecting on the sense of satisfaction that this generates. If a person finds that their achievement leads to a happier mood, they may be ready to put more effort into next challenge to achieve. In the longer term, being able provide positive reinforcement for achievement may lead to more efficient functioning and better self-esteem.

For positive rumination to work, a person would first have to be aware of how they are feeling emotionally, which can be a problem in ADHD. This involves stopping, thinking and developing their self-awareness of their mood. Positive rumination may need to be specifically taught and practiced in order for a person with ADHD to be able to use it effectively and understand and recognise its value.

There is evidence that frequent small, positive emotional boosts are associated with enhanced physical and mental well-being. Positive rumination might provide this, but to be a workable strategy it would depend on the individual taking time to consider the good points about a piece of work or an activity and then reflecting on the sense of satisfaction that is generated. For example, after doing a piece of work, even if the work is not perfect, some positive attributes may be identified. These could initially be pointed out by the parent or teacher, but ultimately the individual would be encouraged to identify for themselves the value in their work. Times of reflection may also be built into the daily routine, for example, at bedtime thinking of the positive and enjoyable experiences and achievements of the day. These might include some of the following pleasant activities that are often associated with positive emotions: being helpful, interactions with others, playing, learning, exercise and spiritual activities. In positive rumination, the individual must be able to pause and reflect and have awareness of their mood, together with mood changes following on from their positive reflection.

Spending time enjoying an achievement lifts the mood

Positive re-appraisal

In therapeutic settings, emotional self-regulation has tended to be directed towards dealing with negative emotions, for example, in anxiety, depression and anger. However, a more recent approach to emotional regulation aims to generate and promote positive emotions. Some strategies, such as negative rumination, avoidance and suppression are associated with psychopathology, while re-appraisal, problem-solving and acceptance are considered protective.

Positive re-appraisal involves redefining an adverse event in terms of the possible positive aspects and actively looking to benefit from experience. For example, when a person is in trouble, if his or her mood and self-esteem can be preserved, the person may be less tempted to resort to bullying in order to feel better. Therefore, they might rationalise the experience and think about what they have learned and how to do better in the future.

Today I said something thoughtless and upset my friend. I felt so bad about it.

Positive re-appraisal involves preserving the mood and self-esteem by looking for something that is good

But then I said sorry, and we made up. I feel better when I've been kind to my friends.

Anger Management

Anger is often a significant problem for people with ADHD and ODD. Anger may consist of a low or irritable baseline mood, or at the other extreme it may involve acute episodes of rage. If a person's mood is lower than it should be, their anger is nearer the surface and more easily triggered. In some people this can be recognised when they show their irritability with a high rate of swearing.

Anger makes it far more difficult for a person to behave rationally. It is as if the anger takes over the decision-making process so that the brain cannot function properly. A person who is angry may feel like releasing the energy, and this may help them to get over it. Anger may be released in a way that causes minimal harm, such as swearing, or punching or kicking an object without damaging it. If the anger is more severe, the person will want to cause hurt, damage or injury. During a rage attack a person becomes so angry that they may lose all control, perhaps afterwards having cause to deeply regret their actions, particularly if they have injured someone. They may also have little recollection after the event.

A person who is susceptible to severe bouts of anger needs to learn to recognise the warning signs, so that they can move away from the situation and avoid causing harm. Signs of anger may include physical tensing, clenching the fists or wanting to shout and swear. Being aware of feeling hot and having a pounding heart may also help with anger recognition. A person can then move away to a safe place where they can let their anger out harmlessly. This might be by screaming and hitting a punching bag or going for a run. Calming strategies can also be useful and may include concentrating on controlled breathing or taking deep breaths with self-instructions such as 'calm down' or 'relax' while breathing out. Imagining reducing the body temperature and heartbeat may also help.

When a person is impulsive due to ADHD and has a low baseline mood and is easily triggered to anger due to ODD, the potential for conflict may be substantial. Such a person may misinterpret other people's actions as hostile and therefore easily become angry if another person's motive is unclear. An example might be impulsively retaliating with violence to an accidental push or shove (see picture on page 41). If questioned afterwards they may try to justify the aggression by blaming the other person.

As people mature, they often realise that they make better choices if they delay making any immediate response while they are feeling angry. Therapy may emphasise thinking of non-personal reasons to explain another's behaviour instead of taking offense (for example: 'the boss might just have been having a bad day').

Emotional self-regulation with the dual aims of promoting a good mood and recognising the signs of anger so that loss of control can be prevented, appears logical and sensible. If strategies can be used effectively by people with ADHD, they could lead to improvements in mood, functioning and self-esteem which would not be linked to specific tasks and situations. The lack of study of emotional regulation in ADHD does not necessarily mean that such strategies are not being used therapeutically and effectively. However, efficacy still needs to be established with further research.

12. Pharmacological management of ADHD

A person with ADHD may learn strategies that help them compensate and get through their daily life. However, only medication can address the underlying problems with the brain's capacity to function efficiently and can also reset the baseline mood.

> ## The aim of treatment:
> ### achieving and being able to function normally

- For a person to learn and remember, they must be able to concentrate
- Control of behaviour may not be readily achievable without stabilising the mood and emotions

If attention is inconsistent, a person with ADHD may have difficulty learning, remembering and organising their tasks, and thinking sufficiently well to make rational choices. They would also experience mental fatigue with tasks that require sustained concentration. The tendency to act quickly and impulsively without the opportunity for adequate decision-making can greatly reduce the efficacy of behavioural management strategies. This is because behaviour management depends on the person being able to make a rational decision based on the pre-determined consequences. Furthermore, the low mood that is associated with reward deficit will tend to reduce the inclination to cooperate. People who have significant functional impairment due to ADHD are sometimes identifiable as those who do not respond to the management strategies that work well for their siblings or peers.

A person whose anger is easily triggered may become better at understanding and managing their anger, but the effort required for this may generate stress. In adults, even the occasional episode of anger getting out of control can have a devasting effect at home or in the workplace. Furthermore, the low mood that is associated with reward deficit will tend to lead to a negative, unco-operative attitude. This may make the person less well-liked.

- It is most important to find the medication and dose that works best.
- Medication is continued for as long as necessary and may be lifelong.

The aim of treatment in ADHD is to enable a person to function efficiently and achieve their goals that are realistic for their level of ability. People who have significant difficulties in coping with life due to ADHD and ODD are sometimes recognisable as those who do not respond to good, consistent management strategies that they may be working on with their therapist or coach. Alternatively, they may be receiving treatment for another condition such as depression or anxiety and not responding well due to their untreated ADHD. Behaviour management strategies can be very helpful, but only medication can target the neurochemistry that underlies ADHD. Therefore, even with the best behaviour management strategies, the mental inefficiencies that are fundamental to ADHD will remain.

Stimulants to improve the symptoms of ADHD and ODD

The medications used most frequently in ADHD are the stimulants. They enhance the levels of neurotransmitters, which are the chemicals that enable communication between the different cells in the brain. This generally results in improvements in the efficiency of the 'thinking' brain. Stimulants also improve the mood and behaviour, which may be an effect of enhancing the activity of the dopamine reward pathway.

The beneficial effect of stimulants in ADHD was first recognised in the 1930s. Since then, numerous trials comparing them with placebo (inactive tablets) have confirmed that the stimulants are effective for treating ADHD. In fact, the stimulants are almost certainly the most studied and the most effective drugs used in psychiatry. They work in pre-schoolers, school aged children, adolescents and adults, including the elderly, improving the ability for sustained attention. They also suppress the appetite. Although usually combined with behavioural interventions, the stimulants often have a more immediate and more obvious effect than behaviour therapy.

In the Multimodal Treatment study of ADHD (MTA Study), it was shown that for the core features of ADHD (inattention, hyperactivity and impulsivity), stimulant medication was more effective than behaviour therapy, while behavioural management was better for comorbid conditions including ODD. Children treated only with stimulant medication required higher doses for improvement than those who used a combination of medication and behaviour therapy, suggesting an interaction between the two strategies.

Side effects
The most significant side effect of using stimulant medication for treating ADHD is usually the effect on appetite and weight. It is as if the stimulant resets the appetite at a lower level. This is a bit like turning down the thermostat when heating a room. The heater goes off and the room cools down until it reaches the temperature where the thermostat is triggered, and the heater starts up again. A person therefore loses weight initially, but after some weeks the appetite picks up and the weight stabilises. Appetite suppression appears to correlate closely with the therapeutic effect. This means that a dose that does not cause any weight loss when used consistently is likely to be too low to be effective.

As the effect of the stimulant wears off later in the day, the appetite returns there may be rebound, with moodiness and irritability. It is as if the behaviour that is held in check by the medication is released as the medication wears off. Rebound may be helped by using a long acting formulation that wears off more gradually, or taking an additional, smaller dose, later in the day.

Stimulant medication also increases the heart rate and blood pressure and can cause difficulty sleeping, irritability and feelings of sadness. The sadness usually improves over the first 2-3 weeks although some people continue to feel lower in their mood while on stimulants. The long-term effects of the higher heart rate and blood pressure are unknown, but it is possible that these effects, if uncontrolled, may increase the risk of having a heart attack or stroke. However, if the blood pressure is being checked regularly, high blood pressure may be detected and treated early.

Stimulants have been associated with tics (habit spasms, such as twitches of the face or eyes, or repeated throat clearing). However, tics are more common in people with ADHD. Tics tend to come and go, getting worse for a few months and then improving. They may therefore coincide with starting stimulant medication. If necessary, medication could be ceased, to find out whether this makes the tic improve.

Treating ADHD with medication is a process of constant reassessment, always looking at the advantages and disadvantages of treating versus not treating. If the medication is working well, some side effects may be tolerated, keeping in mind that no medication is perfect.

Using short or long acting (slow release) medications
The stimulant medications dexamphetamine and methylphenidate are short acting, with an effect that lasts around 3-4 hours. (Methylphenidate is the stimulant used in Ritalin and Concerta). In people who have significant hyperactivity or oppositional symptoms the effect is usually obvious in the first 30 minutes after taking medication. In people who only have inattention, the

effect may be more subtle. Because the stimulants can cause difficulty with settling off to sleep at night, medication is often targeted to be effective earlier in the day, wearing off into the evening. This also allows the appetite to recover so that the person is able to make up for the reduced appetite through the day.

The beneficial effect of stimulant medication may be prolonged by using capsules that release the medication slowly over several hours. Slow release capsules also wear off more slowly, which may reduce the rebound effect. Formulations of methylphenidate include short acting Ritalin tablets (duration 3-4 hours) and the longer acting Ritalin LA (6-8 hours) and Concerta (8-10 hours). Dexamphetamine is also available as lisdexamfetamine (Vyvanse), in which the dexamphetamine has been inactivated by combining it with a protein molecule. It is reactivated in the body, but this process means that it is retained for longer, usually lasting 8-12 hours; it is therefore only taken once daily.

Using a short acting medication can be inconvenient because it is regularly wearing off and the person has to keep remembering to take their next dose. However, there can also be advantages. Firstly, there is always the opportunity for the person and their family to compare their functioning on and off medication. This means that medication can be targeted to the times when it is needed most. However, it is always important for a person with ADHD to take advice from family and workmates because the problems associated with ADHD and improvements on medication are often more obvious to other people.

Choosing the right stimulant and establishing the dose
The stimulants dexamphetamine and methylphenidate are similar in their beneficial effects and their side effects. Most people with ADHD will have a good response to either medication but some definitely do better on one than the other.

The actions of the stimulants are complex in that they have different effects in different parts of the brain and that the optimal doses for different aspects of functioning may not coincide. The dose–response curves for cognitive functioning and for mood and behaviour might look like Figure 5.

Figure 5: Hypothetical dose–response curves for the improvements in executive functioning and behaviour on stimulant medication

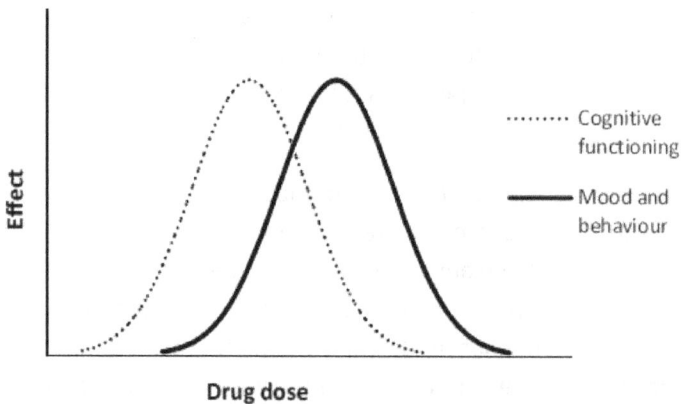

This figure illustrates that the dose can be titrated to maximise either effect, but not both together. Alternatively, the dose giving the best overall effect might fall somewhere between the two peaks. Even if one effect is targeted, the selected dose may still lead to some improvements in the other effect. For example, a child with severe symptoms of ODD may function best on a relatively high dose of medication. Although this dose may be higher than their optimal dose for executive functioning, they may still concentrate substantially better than they would if unmedicated. This is likely to be related to some improvement in his executive functioning on the selected dose and because their attitude towards cognitive tasks may be better when the deficits in their dopamine neurotransmission are addressed.

One of the most important aspect of using these medications is to find the dose that works best for the individual. This is usually done by starting at a low dose and gradually increasing the dose while observing the changes in functioning. As the dose is increased, there is usually progressively more improvement in functioning until a level is reached where further increases do not lead to any further improvement. This is the optimal dose. If the optimal dose is exceeded the behaviour may worsen: some people become angrier; others become withdrawn and depressed. If this happens, the dose should be reduced.

Figure 6: Stimulant medication and the effects of dose titration

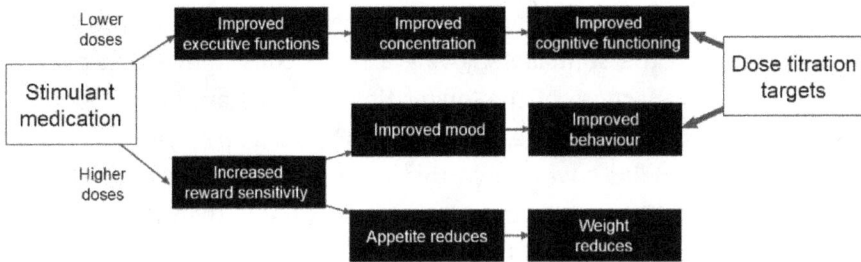

As children mature, they tend to improve in their behaviour. They may also outgrow their dose as they gain weight. Therefore, a dose initially selected for optimal improvement in the symptoms of ODD may, with time, gradually progress into a dose that is better for maximising the executive functioning deficits as the drug levels decline with the growth of the child.

Abuse potential

One ongoing concern about using the stimulants is the risk of abuse and diversion. Although chemically similar to cocaine and methamphetamine ('speed'), the stimulants used in ADHD are far less addictive. This is because they take longer to enter the brain and bind with the dopamine receptors, which makes them less euphoric. This means that people who abuse stimulants are more likely to use them so that they can work or study for longer, rather than for the euphoria. It is reassuring that even though stimulants have been used in ADHD for more than half a century, there is still very little evidence that people treated for ADHD are at risk of becoming addicted to their stimulant.

Other medications used in ADHD

Atomoxetine

Atomoxetine (Strattera) has been developed for treating ADHD. It is not a stimulant and therefore lacks the abuse potential of the stimulants. It is also longer acting than the stimulants, giving a more consistent effect over the course of the day. Because of the longer time that it stays in the body, it is given as a low dose and may take several weeks to build up to give an adequate effect. Although studies have shown that the majority of people with ADHD respond to atomoxetine, the response is more variable than the stimulants. Atomoxetine has been shown to be beneficial for people with ADHD and anxiety.

Clonidine

Clonidine (Catapres) is a medication that can be helpful in ADHD. It can improve the symptoms of ADHD but is usually less effective than the stimulants and may need to be given more frequently to give a consistent effect. Sometimes it is used to prolong the effect of a stimulant or to balance out the side effects, as it causes sleepiness and may increase the appetite. It can also be helpful for anger and aggression. It was developed to treat high

Dr. Alison Poulton
poultonadhd.com.au

blood pressure and it drops the blood pressure and heart rate, which can be a problem in overdose.

Guanfacine

Guanfacine (Intuniv) is similar to clonidine but is longer lasting, requiring only once daily dosing. It is therefore more convenient and also more effective than clonidine.

13. Summary

- ADHD is a common condition which may have a life-long effect on functioning.
- It is important to consider ADHD in adults and children of all ages who appear not to be achieving as they should.
- The longer a person's diagnosis and treatment is delayed, the more they are likely to regret the years they spent struggling with their ADHD.
- The goal of treatment is normal functioning.

The main novel approach to treatment suggested in this book is the recommendation for strategies designed to enhance the positive emotional experiences in everyday life for individuals with ADHD. These would clearly need to be evaluated with randomised controlled studies that include a plausible comparison treatment. In young children, behaviour management strategies are generally taught to the parents who then implement them with the child. Therefore, groups of parents could be taught conventional behaviour management using external rewards and punishments or strategies designed to promote positive emotions in the child through their achievement. Outcomes would be assessed using standardised rating scales relating not only to achievement in terms of task completion but also any positive effects on mood. In older children and adults with ADHD, particularly those with anger or oppositional features, there would be value in comparing anger management strategies that are intended to give more control over negative emotions, with strategies designed to enhance the positive experience of reward. Outcomes could be evaluated with standardised rating scales, both self-reported and observer-reported.

References

1. Introduction – What is ADHD?

- Amiri S, Fakhari A, Maheri M, Mohammadpoor Asl A. Attention deficit/hyperactivity disorder in primary school children of Tabriz, North-West Iran. Paediatr Perinat Epidemiol 2010; 24 597-601.
- Dopfner M, Breuer D, Wille N, Erhart M, Ravens-Sieberer U. How often do children meet ICD-10/DSM-IV criteria of attention deficit-/hyperactivity disorder and hyperkinetic disorder? Parent-based prevalence rates in a national sample – results of the BELLA study. Eur Child Adolesc Psychiatry 2008; 17 Suppl 1 59-70.
- Sawyer MG, Arney FM, Baghurst PA, Clark JJ, Graetz BW, Kosky RJ, Nurcombe B, Patton GC, Prior MR, Raphael B et al. The mental health of young people in Australia: key findings from the child and adolescent component of the national survey of mental health and well-being. Aust NZ J Psychiatry 2001; 35 806-14.
- Caci HM, Morin AJ, Tran A. Prevalence and correlates of attention deficit hyperactivity disorder in adults from a French community sample. J Nerv Ment Dis 2014; 202 324-32.
- Antshel KM, Faraone SV, Maglione K, Doyle A, Fried R, Seidman L, Biederman J. Is adult attention deficit hyperactivity disorder a valid diagnosis in the presence of high IQ? Psychol Med 2009; 39 1325-35.
- de Zwaan M, Gruss B, Muller A, Graap H, Martin A, Glaesmer H, Hilbert A, Philipsen A. The estimated prevalence and correlates of adult ADHD in a German community sample. Eur Arch Psychiatry Clin Neurosci 2012; 262 79-86
- American Psychiatric Association DSM Task Force. Diagnostic and Statistical Manual of Mental Disorders: DSM-5. Arlington, VA: American Psychiatric Association; 2013.
- Rubia K, Smith AB, Halari R, Matsukura F, Mohammad M, Taylor E, Brammer MJ. Disorder-specific dissociation of orbitofrontal dysfunction in boys with pure conduct disorder during reward and ventrolateral prefrontal dysfunction in boys with pure ADHD during sustained attention. Am J Psychiatry 2009; 166 83-94.
- Pennington BF, Ozonoff S. Executive functions and developmental psychopathology. J Child Psychol Psychiatry 1996; 37 51-87.
- Del Campo N, Chamberlain SR, Sahakian BJ, Robbins TW. The roles of dopamine and noradrenaline in the pathophysiology and treatment of attention-deficit/hyperactivity disorder. Biol Psychiatry 2011; 69 e145-57

- Hulvershorn LA, Mennes M, Castellanos FX, Di Martino A, Milham MP, Hummer TA, Roy AK. Abnormal amygdala functional connectivity associated with emotional lability in children with attention-deficit/hyperactivity disorder. J Am Acad Child Adolesc Psychiatry 2014; 53 351-61.e1.
- Hoogman M, Bralten J, Hibar DP, Mennes M, Zwiers MP, Schweren LS et al. Subcortical brain volume differences in participants with attention deficit hyperactivity disorder in children and adults: a cross-sectional mega-analysis. The lancet Psychiatry. 2017. doi:10.1016/s2215-0366(17)30049-4.
- Gamo NJ, Wang M, Arnsten AF. Methylphenidate and atomoxetine enhance prefrontal function through alpha2-adrenergic and dopamine D1 receptors. J Am Acad Child Adolesc Psychiatry 2010; 49 1011-23.
- Sinita E, Coghill D. The use of stimulant medications for non-core aspects of ADHD and in other disorders. Neuropharmacology 2014; 87: 161-72.
- Biederman J, Mick E, Faraone S. Age-dependent decline of symptoms of attention deficit hyperactivity disorder: impact of remission definition and symptom type. Am J Psychiatry 2000; 157 816-8.

5. Living with ADHD across the lifespan

- Franke B, Michelini G, Asherson P, et al. Live fast, die young? A review on the developmental trajectories of ADHD across the lifespan. Eur Neuropsychopharmacol 2018; 28(10): 1059-88.
- Zalsman G, Shilton T. Adult ADHD: A new disease? International journal of psychiatry in clinical practice 2016; 20(2): 70-6.

6. Oppositional Defiant Disorder (ODD) / Antisocial Personality Disorder

- Stringaris A, Goodman R. Mood lability and psychopathology in youth. Psychol Med 2009; 39 1237-45.
- Fondacaro MR, Heller K. Attributional style in aggressive adolescent boys. J Abnorm Child Psychol 1990; 18 75-89.

7. Mental Effort-Reward Imbalances Model (MERIM): Why do people with ADHD so often have ODD as well?

- Poulton A, Nanan R. The attention deficit hyperactivity disorder phenotype as a summation of deficits in executive functioning and reward sensitivity: does this explain its relationship with oppositional defiant disorder? Australas Psychiatry 2014; 22 174-8.
- Poulton A. Therapy for ADHD directed towards addressing the dual imbalances in mental effort and reward as illustrated in the Mental Effort-Reward Imbalances Model (MERIM) In Norvilitis JM editor: ADHD - New Directions in Diagnosis and Treatment. Intechopen.com 2015. ISBN: 978-953-51-2166-4. http://www.intechopen.com/download/pdf/48642
- Keyes CL. Promoting and protecting mental health as flourishing: a complementary strategy for improving national mental health. Am Psychol 2007; 62 95-108.
- Catalino LI, Fredrickson BL. A Tuesday in the life of a flourisher: the role of positive emotional reactivity in optimal mental health. Emotion 2011; 11 938-50.
- Downs B, Oscar-Berman M, Waite R, Madigan M, Giordano J, Beley T, Jones S, Simpatico T, Hauser M, Borsten J et al. Have We Hatched the Addiction Egg: Reward Deficiency Syndrome Solution System. Journal of genetic syndrome & gene therapy 2013; 4 14318.
- Barkley RA. Behavioral inhibition, sustained attention, and executive functions: constructing a unifying theory of ADHD. Psychol Bull 1997; 121 65-94.
- Parry PA, Douglas VI. Effects of reinforcement on concept identification in hyperactive children. J Abnorm Child Psychol 1983; 11 327-40.

8. ADHD as a continuum

- Pauli-Pott U, Neidhard J, Heinzel-Gutenbrunner M, Becker K. On the link between attention deficit/hyperactivity disorder and obesity: do comorbid oppositional defiant and conduct disorder matter? Eur Child Adolesc Psychiatry 2014; 23 531-7.
- Mustillo S, Worthman C, Erkanli A, Keeler G, Angold A, Costello EJ. Obesity and psychiatric disorder: developmental trajectories. Pediatrics 2003; 111 851-9.
- Harty SC, Ivanov I, Newcorn JH, Halperin JM. The impact of conduct disorder and stimulant medication on later substance use in an ethnically diverse sample

of individuals with attention-deficit/hyperactivity disorder in childhood. J Child Adolesc Psychopharmacol 2011; 21 331-9.
- Modesto-Lowe V, Danforth JS, Neering C, Easton C. Can we prevent smoking in children with ADHD: a review of the literature. Conn Med 2010; 74 229-36.

10. ADHD and mental illness

- Katzman MA, Bilkey TS, Chokka PR, Fallu A, Klassen LJ. Adult ADHD and comorbid disorders: clinical implications of a dimensional approach. BMC Psychiatry 2017; 17(1): 302.
- Cortese S, Vincenzi B. Obesity and ADHD: Clinical and Neurobiological Implications. Curr Top Behav Neurosci 2012; 9: 199-218.
- Evren B, Evren C, Dalbudak E, Topcu M, Kutlu N. Relationship of internet addiction severity with probable ADHD and difficulties in emotion regulation among young adults. Psychiatry Res 2018; 269: 494-500.

11. Non-pharmacological management of ADHD

- Cohn MA, Fredrickson BL. In search of durable positive psychology interventions: Predictors and consequences of long-term positive behavior change. The journal of positive psychology 2010; 5 355-66.
- Nolen-Hoeksema S, Wisco BE, Lyubomirsky S. Rethinking Rumination. Perspectives on Psychological Science 2008; 3 400-24.
- McLaughlin KA, Aldao A, Wisco BE, Hilt LM. Rumination as a transdiagnostic factor underlying transitions between internalizing symptoms and aggressive behavior in early adolescents. J Abnorm Psychol 2014; 123 13-23.
- Conway F. The use of empathy and transference as interventions in psychotherapy with attention deficit hyperactive disorder latency-aged boys. Psychotherapy (Chic) 2014; 51 104-9.
- Aldao A, Nolen-Hoeksema S, Schweizer S. Emotion-regulation strategies across psychopathology: A meta-analytic review. Clin Psychol Rev 2010; 30 217-37.
- Fuller JR, Digiuseppe R, O'Leary S, Fountain T, Lang C. An open trial of a comprehensive anger treatment program on an outpatient sample. Behav Cogn Psychother 2010; 38 485-90.
- Feindler EL. Ideal treatment package for children and adolescents with anger disorders. Issues Compr Pediatr Nurs 1995; 18 233-60.

- Klein SA, Deffenbacher JL. Relaxation and exercise for hyperactive impulsive children. Percept Mot Skills 1977; 45 1159-62.

12. Pharmacological management of ADHD

- MTA Cooperative Group. A 14-month randomized clinical trial of treatment strategies for attention-deficit/hyperactivity disorder Arch Gen Psychiatry 1999; 56 1073-86.
- Rapport MD, Denney C. Titrating methylphenidate in children with attention-deficit/ hyperactivity disorder: is body mass predictive of clinical response? J Am Acad Child Adolesc Psychiatry 1997; 36 523-30.
- Bradley C. The behavior of children receiving benzedrine. Am J Psychiatry 1937; 94 577-85.
- Greenhill LL, Halperin JM, Abikoff H. Stimulant medications. J Am Acad Child Adolesc Psychiatry 1999; 38 503-12.
- Spencer TJ, Abikoff HB, Connor DF, Biederman J, Pliszka SR, Boellner S, Read SC, Pratt R. Efficacy and safety of mixed amphetamine salts extended release (adderall XR) in the management of oppositional defiant disorder with or without comorbid attention-deficit/hyperactivity disorder in school-aged children and adolescents: A 4- week, multicenter, randomized, double-blind, parallel-group, placebo-controlled, forced-dose-escalation study. Clin Ther 2006; 28 402-18.
- Greenhill L, Kollins S, Abikoff H, McCracken J, Riddle M, Swanson J, McGough J, Wigal S, Wigal T, Vitiello B. Efficacy and safety of immediate-release methylphenidate treatment for preschoolers with ADHD. J Am Acad Child Adolesc Psychiatry 2006; 45 1284-93.
- Jain U, Hechtman L, Weiss M, Ahmed TS, Reiz JL, Donnelly GA, Harsanyi Z, Darke AC. Efficacy of a novel biphasic controlled-release methylphenidate formula in adults with attention-deficit/hyperactivity disorder: results of a double-blind, placebo-controlled crossover study. J Clin Psychiatry 2007; 68 268-77.
- Blader JC, Pliszka SR, Jensen PS, Schooler NR, Kafantaris V. Stimulant-responsive and stimulant-refractory aggressive behavior among children with ADHD. Pediatrics 2010; 126 e796-806.
- Newcorn JH, Spencer TJ, Biederman J, Milton DR, Michelson D. Atomoxetine treatment in children and adolescents with attention-deficit/hyperactivity disorder and comorbid oppositional defiant disorder. J Am Acad Child Adolesc Psychiatry 2005; 44 240-8.

Dr. Alison Poulton
poultonadhd.com.au

- Sprague RL, Sleator EK. Methylphenidate in hyperkinetic children: differences in dose effects on learning and social behavior. Science 1977; 198 1274-6.
- Volkow ND, Wang GJ, Fowler JS, Gatley SJ, Logan J, Ding YS, Hitzemann R, Pappas N. Dopamine transporter occupancies in the human brain induced by therapeutic doses of oral methylphenidate. Am J Psychiatry 1998; 155 1325-31.
- Poulton A, Cowell CT. Slowing of growth in height and weight on stimulants: a characteristic pattern. J Paediatr Child Health 2003; 39 180-5.
- Stiefel G, Besag FM. Cardiovascular effects of methylphenidate, amphetamines and atomoxetine in the treatment of attention-deficit hyperactivity disorder. Drug Saf 2010; 33 821-42.
- Efron D, Jarman F, Barker M. Side effects of methylphenidate and dexamphetamine in children with attention deficit hyperactivity disorder: a double-blind, crossover trial. Pediatrics 1997; 100 662-6.
- Cortese S, Adamo N, Del Giovane C, et al. Comparative efficacy and tolerability of medications for attention-deficit hyperactivity disorder in children, adolescents, and adults: a systematic review and network meta-analysis. The lancet Psychiatry 2018.
- Huss M, Dirks B, Gu J, Robertson B, Newcorn JH, Ramos-Quiroga JA. Long-term safety and efficacy of guanfacine extended release in children and adolescents with ADHD. Eur Child Adolesc Psychiatry 2018; 27(10): 1283-94.

www.ingramcontent.com/pod-product-compliance
Lightning Source LLC
Chambersburg PA
CBHW072153020426
42334CB00018B/1986